MICROCIRCUIT LEARNING COMPUTERS

The Learning Computer is quite different in concept from a conventional computer. It consists of vast numbers of microcircuit learning elements connected in a network, the aim of which is to process data not by program but by example. In a way, it resembles the human brain which is also a network of learning elements: neurons.

This monograph reviews the recent work done on the subject (mainly within the author's research group) and aims to provide a foundation for those who would like to pursue the topic further. Chapter 1 describes the broad concepts of the nature of learning elements, networks and applications. Chapter 2 details the design of digital learning elements, describing in particular the SLAM (Stored Logic Adaptive Microcircuit) module, which is the building brick of learning computers. Chapter 3 presents analyses of the learning behaviour of networks of SLAMs while Chapter 4 describes their application to pattern recognition and automatic control. Chapter 5 reviews the most recent work on learning networks with feedback used for the processing of sequences of patterns.

Other titles in this series

EE/1 THRESHOLD LOGIC
S L Hurst

EE/2 THE MAGNETRON OSCILLATOR
E Kettlewell

EE/3 MATERIAL FOR THE GUNN EFFECT
J W Orton

EE/5 D C CONDUCTION IN THIN FILMS
J G Simmons

Other Electrical Engineering titles available

TL/EE/1 HYDROSTATIC EXTRUSION
J M Alexander and B Lengyel

TL/EE/2 AN INTRODUCTION TO THE JOSEPHSON EFFECTS
B W Petley

All published by Mills and Boon Limited.

M & B Monograph EE/4

General Editor: J Gordon Cook, PhD FRIC

Microcircuit Learning Computers

I Aleksander, PhD
Reader in Electronics
University of Kent

Mills & Boon Limited
London

First published in Great Britain 1971
by Mills & Boon Limited, 17–19 Foley Street,
London, W1A 1DR

© Mills & Boon Limited 1971
ISBN 0.263.51724.1

All rights reserved. No part of this publication
may be reproduced, stored in a retrieval system,
or transmitted in any form or by any means,
electronic, mechanical, photocopying, recording
or otherwise, without the prior permission of
Mills & Boon Limited.

Printed by photo-lithography and
made in Great Britain at the Pitman Press, Bath

CONTENTS

1. INTRODUCTION TO ELECTRONIC LEARNING	9
Learning	9
Generalisation	12
Combinational Applications	14
Feedback: the "Thought" Mechanism of Learning Computers	16
2. LOGIC LEARNING ELEMENTS	18
Universal Logic Circuits	19
Control Memory Organisation: Function Searching	25
Bit Addressed Memory: SLAM Elements	27
3. NETWORKS OF LEARNING ELEMENTS	32
The Generalisation Factor G.	33
Hamming Distance Analysis	35
Comment on Other Topologies	39
4. APPLICATIONS	41
Recognition of Hand-Written Numerals	41
A Note on Feature Extraction	47
Automatic Control	50
5. LEARNING NETS WITH FEEDBACK	53
Improved Recognition	54
Short-Term Memory	61
Recognition of Sequences	64
Recall of Sequences	67
Attention	71
Bibliography	75
Index	77

PREFACE

Since about 1965 it has been apparent that digital microcircuits have far greater potential than is required by the conventional digital computer. I was attracted by the idea that large scale integration (LSI) would make possible developments in learning machines that were hitherto prohibited by the high cost of discrete-component electronics. As a result, SLAM (Stored Logic Adaptive Microcircuit) elements were developed and systems of such elements were investigated. In this monograph, the developments that have taken place are described. This has been done by reviewing the papers which have been published, and by extracting from them what now appear to be the fundamentals of the digital learning computer.

We are only at the beginning of this development in electronics. It is hoped that those who read this monograph will have their curiosity aroused enough to make them realise that the field is wide open and ready to welcome the fruits of imaginative thought.

I. A. September, 1970.

1. Introduction to Electronic Learning

This monograph is concerned primarily with learning machines that are *electronic* in nature. We shall not concern ourselves with general ideas of mechanised learning, as this topic is covered adequately by other books (Nilsson, 1965; Michie, 1967–1970). The first chapter will be devoted, therefore, to an explanation of the salient *electronic* issues involved in learning systems. This will clarify the reasons for devoting attention to various forms of detail in the later chapters.

The subject is introduced by qualifying what is meant by learning in an electronic system. This is followed by an introduction to an aspect of the *quality* of learning: *generalisation*. An outline of the *applications* of simple learning systems is then given. These lie mainly in the fields of pattern recognition and automatic control.

The most intriguing modes of behaviour in learning systems occur when a simple electronic learning net is provided with *feedback loops*. These are discussed at the end of this chapter.

LEARNING

It is important to establish what is meant by the word *learning* when applied to an electronic system. Psychologists define learning as:

" ... a change in behaviour as a result of experience ... "

Can one say therefore, if one changes the electrical behaviour of a variable resistor by changing its setting, that the resistor has learned anything? Intuitively, one feels that it would be trivial to define electronic learning in this way.

Let us try another example, again involving the setting of a variable resistor. This time, the resistor is used in an analogue computer circuit and connected to a motor as shown in Fig. 1. Matters can be arranged so that the resistor is driven to find a value that will make system A behave as much as possible like system B. The condition for it to do so is that at all times the signal transmitted to the motor should drive it in such a way as to minimise the discrepancy between the two systems. There is now no objection to stating that system A is *learning* to behave like system B. Indeed, the motor-driven variable resistor has played a major role in some electronic learning systems.

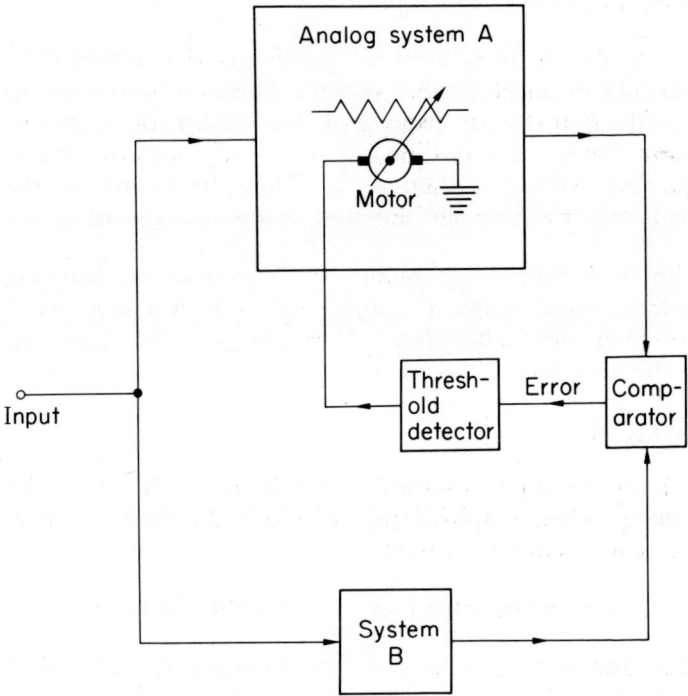

Fig. 1   Variable Resistor Learning System

At this stage we are content to note the slight but crucial difference between the first example of setting the variable resistor and the one in Fig. 1. There are two reasons for which we are satisfied to say that the latter learns and the former does not.

Firstly, the resistor in Fig. 1 may be set automatically, that is, *electrically* in this case. In other words, we have an *electrical* element in an *electrical* system which is sensitive to the *electrical* information which indicates that a change in behaviour is needed. Secondly, the information received by the learning element must, in some way, relate to the desired output of the element. If, for example, the motor in Fig. 1 were made to go in the direction which increases the discrepancy, the system would oscillate or saturate, and nothing would be learnt at all.

There is an important additional property which is common to both examples. This is the property of information storage; the change in behaviour of the element must be retained even after the driving signal has been removed. This is the property that made us select variable resistors as examples of learning elements in the first instance. The resistor is an analogue storage element, and an information store of some kind is necessary in all learning elements. Thus, we have arrived at a simple definition of an electronic learning element:

An electronic learning element is one which changes its electronic behaviour for as long as its use requires, in response to an electronic signal which contains information regarding the desired change in behaviour.

Variable resistor learning elements have, in the past, stemmed from a desire to model neurons of the brain (Aleksander, 1970a,b). The variable resistor plays the role of the variable synaptic weights in which the property of learning is presumed to reside. It is likely that in trying to model analogue-memory neurons, workers in this field have neglected digital techniques of information storage. Chapter 2 is devoted, therefore, to a description of various *digital* learning elements in which the link with neurons has been largely severed, the elements being optimised within the frame of reference of digital microcircuit technology.

GENERALIZATION

It is becoming evident that in referring to the *behaviour* of an electronic system one is implying the relationship between the input and the output of some "black box" of electronics. In the simple example of the variable resistor, *behaviour* is the relationship between voltage and current, either of which could be considered as input or output.

In this loose definition of behaviour one must take care not to exclude *sequential* behaviour. That is, the output may depend not only on the immediate input but also on the past history of the inputs. Also, the output could be a sequence of signals triggered by some inputs. Nevertheless, the *combinational* (that is, non-sequential) box is very important. Sequential behaviour can always be brought about by the addition of feedback loops to a combinational box.* A general model of a combinational learning box is shown in Fig. 2. With the aid of this figure we can standardise some of the terminology used in the rest of this book.

There are two sets of "sensory" input channels which we define as the *pattern* and the *identity* channels. The *identity* channels receive the information regarding the desired output that is to be associated with the *pattern* currently at

---

* These and other concepts pertaining to non-learning logic circuits are explained in "An introduction to logic circuit theory" (Aleksander, 1970c).

the input. In *training*, or *teaching*, both inputs are energised, while *in use*, or *testing*, only the *pattern* inputs are active. In the latter case the output is the system's computed response to the pattern, that is, the computed identity.

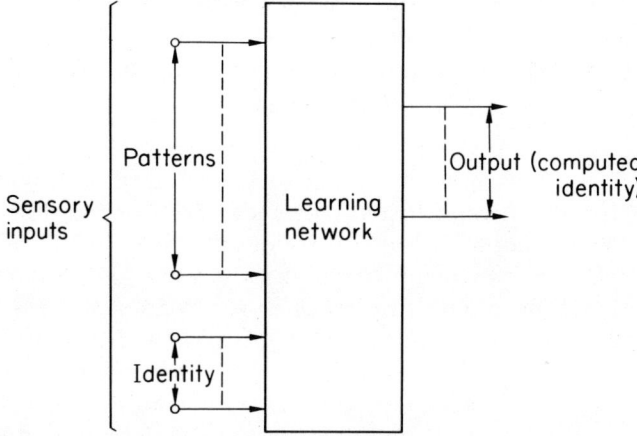

Fig. 2   General Combinational Learning Net

Generalisation is the ability to respond correctly to input patterns not seen during training. This is not a helpful definition unless we can say what we mean by *correctly*. One could say that a correct response implies the presence of input patterns *similar* to those seen in training. But who defines *similarity*? As far as possible this must be defined by the pattern-identity pairs seen during training. For example, a system trained to respond correctly to patterns of capital, hand-written letters cannot be expected to respond correctly to cursive hand-writing; neither could a human, in fact. A fundamental rule of all learning systems is that generalisation is only possible if a representative sample is seen during training.

Technologically, generalisation has an additional implication. A system which does not generalise must be trained to cope with all eventualities. This implies not only long training sessions, but also large amounts of storage to absorb the information. Clearly, the way one tackles this problem electronically is of vital importance. Chapter 3, therefore, is concerned mainly with the effects of network structure on system generalisation. Of particular interest is the possibility of varying the generalisation of a learning system. This is necessary because generalisation is problem-dependent, and a system which over-generalises merely causes an operational error.

COMBINATIONAL APPLICATIONS

Electronic learning systems provide intriguing problems for the electronicist, but this is not enough to justify their existence. Such systems can make a positive and practical contribution to data processing in its broadest sense. It is possible that such systems are of greater fundamental importance and have a wider range of applications than conventional digital computers. The reason is simple: learning systems are designed to deal with non-analytic problems; that is, problems for which it is difficult to provide a mathematical model. The philosopy of learning systems is outlined in the example which follows.

Some of the most erudite mathematical modelling techniques are used in the control of a rocket at lift-off. This takes the form of a control algorithm which is fed to a computer which, in turn, is used as an on-line controller. A broom standing upright on its handle end has dynamic properties similar to those of the rocket, and yet it can be balanced by a child who may have no knowledge of mathematical modelling. The child has *learnt* to control the broom either under the guidance of a teacher or by trial-and-error. This is precisely the mechanism that one wishes to build into a learning system. Before dealing with other examples, let us consider a powerful motive.

In the case of humans or animals there is no way of discovering what it is that has been learnt. Neurophysiologists are making some progress towards discovering the nature of physical changes brought about by learning. However, even if the humanitarian problems could be overcome, one could not derive the human's broom-balancing (or any other) policy by dissecting the brain. In a learning system this presents no difficulty at all. The content of the learning machine's store can be extracted and recorded permanently, say on punched paper tape. The record contains all the information necessary to design a fixed, non-learning controller that performs the learnt task. This could, in fact, be done automatically in a non-learning computer. Therefore, except for special conditions, learning systems will not control processes on-line, but will give birth to very cheap, simple, machines that perform specialised on-line tasks.

The best known, and yet most intractable problem of a non-analytical nature is that of pattern recognition. This is normally tackled in a specialised form, such as the recognition of hand-written characters. Its implications, however, are much wider, particularly in the field of automatic control. The measurements taken on a chemical plant, or the position and angle of the rocket on take-off, are patterns that must be translated into control actions. In the science of information retrieval one finds other examples of pattern recognition, the information to be retrieved being the identity information while given descriptors are the patterns. For example, in medical information retrieval "common cold" is a recognition of the symptom patterns "stuffy nose," "sore throat," "headache" and "normal temperature." Change the last item to "high temperature" and the correct classification becomes: "virus influenza." This is analogous to changing Q to O by the removal of the "squiggle" in character recognition.

Thus pattern recognition is of fundamental importance as a test of a learning system. Chapter 4 is concerned, therefore, with the relationship between the internal connections of a learning system and its behaviour as a pattern recogniser. Results obtained with hand-written characters are used for purposes of illustration. This chapter also deals with other applications, such as the use of learning systems in automatic control. Use is made of the state-variable representation of plant. These variables provide a full description of the dynamics (or sequential behaviour) of the plant and are therefore useful patterns to a net, the output of which provides the controlling action.

In conventional controller design there is no guaranteed solution of the equations that contain these variables: the state equations. In fact, there is no guarantee that such a set of equations may be derived from measurements of state variables on a plant. And yet, the solution of these equations is a necessary prerequisite to the design of an automatic controller. This appears, therefore, to be an ideal application for a learning machine where the state variables are used as pattern information for the machine, while the correct control is obtained as identity information from an experienced controller (possibly human). Trial-and-error learning is also possible on a simulated plant. (It may be disastrous on real plant.) Identity information would have to be derived, in this case, from the success, or lack thereof, of obtaining some prespecified control aim (such as keeping the broom at an inclination of less than one degree to the vertical).

FEEDBACK: THE "THOUGHT" MECHANISM OF LEARNING COMPUTERS

As mentioned earlier, it may be restricting to assume that learning nets must of necessity be combinational. It is standard practice in non-learning logic systems to design sequential (time sensitive) machines by the addition of feedback loops to combinational systems (Aleksander, 1970c). Indeed, the addition of feedback to learning nets has precisely the effect of endowing temporal behaviour to the learning system. This means that such nets capable not only of learning to respond to sequences of input patterns, but they have an interesting autonomous behaviour in that they act as highly non-linear oscillators.

The feedback is applied between the response output of the net and the pattern inputs. Thus the sequences of patterns arising during an "oscillation" are *internal* to the net. An analogy may thus be set up between these oscillations and what we call "thought." Indeed, it has been possible to demonstrate a number of phenomena normally associated with human psychology. These are considered in Chapter 5, where concepts of sequence recall, "wilful" oscillation control and short-term memory are considered among other properties of feedback nets.

2. Logic Learning Elements

Most learning computers designed before the advent of integrated circuit techniques were modelled on *neurons,* the elemental cells of living brains. In this monograph we shall not discuss such devices, since their physical implementation has been singularly disappointing largely due to their need for an analogue store. We shall return briefly to this topic at the end of this chapter, but more details on comparisons with neural nets may be found in Aleksander, 1968a and 1970a. Here, we consider the principles of learning elements that are purely *digital* in nature. This characteristic has led to the manufacture of some of these elements in microcircuit form.

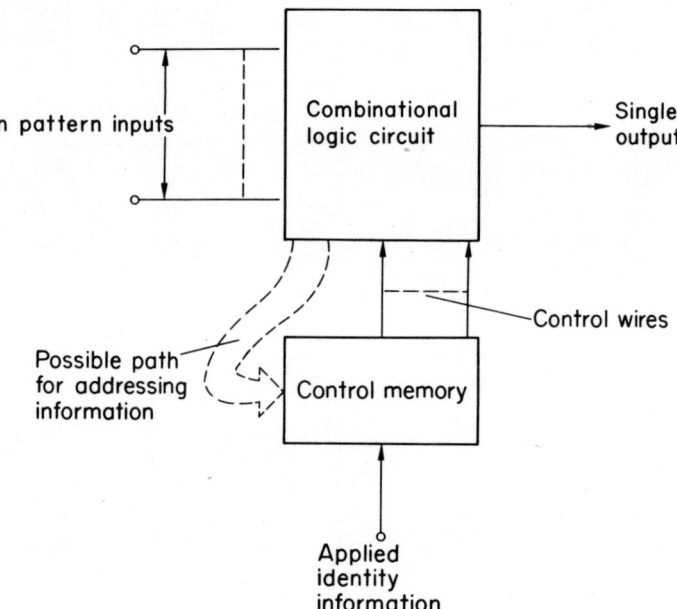

Fig. 3   Learning Element Components

A general diagram of a digital learning element is shown in Fig. 3. The *combinational logic circuit* performs a Boolean operation on the input and the content of the *control memory* to generate the output. This output contributes to the eventual response of the machine which is the *computed identity* of the input. During training, identity information is fed to the *control memory*, and various ways of doing this are discussed here. In some very important elements (c.f. SLAMs) the control memory may also be addressed from the logic circuitry itself. In the first instance, one does not wish to limit the function of the element in any way, and to this end the combinational logic is made *universal*. This means that the element can perform any one of the possible 2^{2^n} functions of its n inputs. The content of the control memory determines *which* of these functions is being executed at any one time.

Here, we limit our discussions to single-output learning elements, noting that an m-output element may be made by connecting m single-output elements in parallel to the n inputs. If all the m elements are universal, the m-output complex performs all the possible 2^{m2^n} functions, and is universal in itself. In this chapter we describe first, the general nature of universal logic systems and second, various ways of organising the control memory, including a description of SLAM, the building brick of other systems considered in this monograph.

UNIVERSAL LOGIC CIRCUITS

An n-input universal logic circuit (u.l.c.) must be able to perform 2^{2^n} Boolean functions. This implies that the control memory, via a set of control wires $\phi = \{\phi_1 \ldots \phi_k\}$ must supply 2^{2^n} messages—one for each function. Thus, there must be 2^n of these ϕ wires. Without loss of generality these may be made to correspond to the entries in a truth table for the function. Then all that is necessary is that they should gate the appropriate minterms or maxterms (see Aleksander, 1970c) to the output. A formal statement of how this is done follows below.

We label the external terminals of the u.l.c. as follows:

Input terminals: $x_0, x_1, \ldots x_{n-1}$
Control terminals: $\phi_0, \phi_1, \ldots \phi_{2^n-1}$
Output terminal: F

One way of defining the relationship between these terminals is by means of the following logic expression:

$$F = \sum_{j=0}^{2^n-1} \phi_j \cdot \left(\prod_{p=0}^{n-1} x_p^{e_p} \right) \qquad (1)$$

where $\sum$ refers to an OR summation
. refers to an AND operation
Π refers to an AND product

$(e_0, e_1 \ldots e_{n-1})$ is an n tuple representing the binary value of j; e.g. if $n = 4$ and $j = 13$, $(e_0, e_1, e_2, e_3) = (1, 1, 0, 1)$
also $x_p^{e_p} = \bar{x}_p$ if $e_p = 0$
and $x_p^{e_p} = x_p$ if $e_p = 1$

We now define three ways of realising the above function:

(a) the 3-level form
(b) the $(2n + 1)$-level form
(c) a t-level form, where $3 < t < (2n + 1)$

Method (a). The 3-level form shown in Fig. 4 arises from a direct implementation of Eqn. 1, and consists of three levels of logic gates. Gate 3 is an OR gate with 2^n inputs, which performs the summing action

$$\sum_{j=0}^{2^n-1}$$

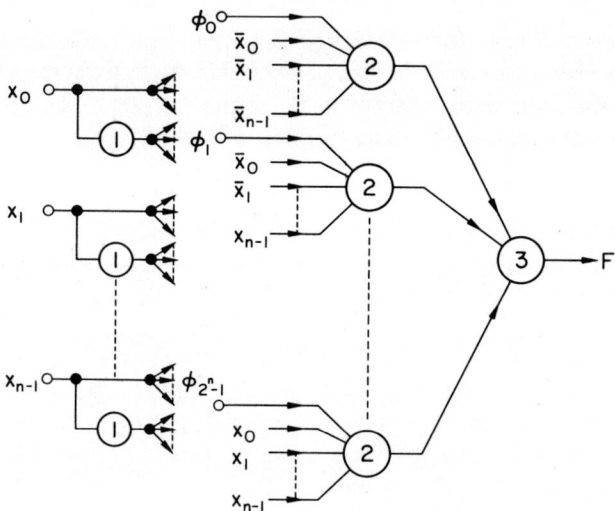

Fig. 4   3-level Universal Logic Circuit

There are 2^n AND gates (gates 2) which provide the functions

$$\phi_j \cdot \prod_{p=0}^{n-1} x_p^{e_p}$$

Finally, gates 1 are NOT elements, which provide the negated versions of the input variables. This function can also be realised by using a NAND element for each of the gates in Fig. 4. There is a dual form of this scheme, which may be obtained by implementing the expression

$$F = \prod_{j=0}^{2^n-1} \left\{ \phi_j + \left(\sum_{p=0}^{n-1} x_p^{\bar{e}_p} \right) \right\} \qquad (2)$$

where $+$ reads OR and $(\bar{e}_0, \bar{e}_1, \ldots \bar{e}_{n-1})$ is the $(e_0, e_1, \ldots e_{n-1})$ n tuple with the 0s replaced by 1s and vice versa. The dual circuit may also be composed entirely of NOR elements. We note that method (a) has the least possible number of gating levels between the inputs and the output, but requires a large fan-in for individual gates; namely, 2^n for gate 3 and n for each gate 2. Scheme (b) is at the opposite extreme—it requires $(2n + 1)$ gating levels, each gate having a fan-in of only 2.

Method (b). The $(2n + 1)$-level type of u.l.c. is shown in Fig. 5. This is obtained by factoring the system in method (a) as far as possible. Analytically, F in this configuration is given by the iterative formula

$$F = (f_0^0) \cdot \bar{x}_0 + (f_1^0) \cdot x_0$$
$$\text{where } (f_0^0) = (f_0^1) \cdot \bar{x}_1 = (f_1^1) \cdot x_1$$
$$\text{and } (f_1^0) = (f_2^1) \cdot \bar{x}_1 + (f_3^1) \cdot x_1$$

and so on, iteratively, with

$$(f_i^k) = (f_{2i}^{k+1}) \cdot \bar{x}_{k+1} + (f_{2i+1}^{k+1}) \cdot x_{k+1}$$

where k is an integer such that $0 < k < n - 1$ and i is an integer such that $0 < i < 2^k - 1$.

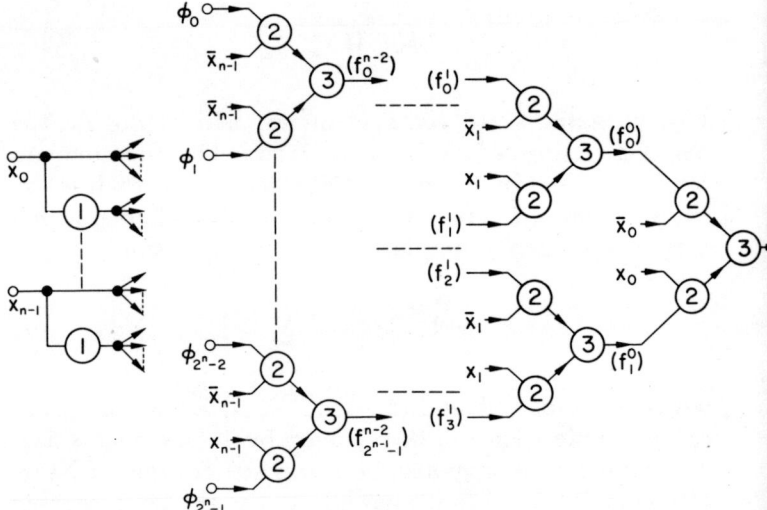

Fig. 5 $(2n+1)$-*level Universal Logic Circuit*

Also, and this is vitally important, we have

$$(f_i^{n-1}) = \phi_i$$

It may be shown by repeated substitution of the iterative steps that Eqn. 2 is indeed equivalent to Eqn. 1. Elements 3, 2 and 1 are either OR, AND and NOT gates, respectively, or all NAND gates. This form of circuit again has a dual in which the OR and AND gates are interchanged and $(f_i^{n-1}) = \phi_{n-i-1}$. This dual may also be composed entirely of NOR elements. We note that the circuits contain $2n$ gating levels preceded by a level of invertors—a total of $(2n + 1)$ levels.

Before discussing method (c), we show a comparison between methods (a) and (b) in Table 1. Method (b) has an advantage over method (a) which has not been mentioned so far. Intermediate outputs may be taken from the circuit, so that an element with n inputs could be used as two universal logic circuits, each producing a different function of $(n - 1)$ variables, or, in general, providing 2^r functions of $(n - r)$ variables.

Table 1

	METHOD (a)	METHOD (b)
Number of gating levels	3	$2n + 1$
Maximum fan-in	2^n	2
Total number of gates (including inventors)	$2^n + n + 1$	$3(2^n - 1) + n$
Total number of inputs	$2^n(n + 2) + n$	$6(2^n - 1) + n$

Method (c). This is a compromise between methods (a) and (b), in which the input variables are grouped, and each group is treated as a separate circuit of type (a). The groups are interconnected in a manner similar to that used in type (b) circuits. To illustrate this method, we consider a particular example in preference to a general treatment. Taking a 4-variable circuit, we investigate three groupings:

Grouping (i): $\{x_0\}\{x_1, x_2, x_3\}$

$$F = (f_0^0) \cdot \bar{x}_0 + (f_1^0) \cdot x_0$$

while

$$(f_0^0) = \phi_0 \cdot \bar{x}_1 \cdot \bar{x}_2 \cdot \bar{x}_3 + \phi_1 \cdot \bar{x}_1 \cdot \bar{x}_2 \cdot x_3 + \ldots \\ + \phi_7 \cdot x_1 \cdot x_2 \cdot x_3$$

and

$$(f_1^0) = \phi_8 \cdot \bar{x}_1 \cdot \bar{x}_2 \cdot \bar{x}_3 + \phi_9 \cdot \bar{x}_1 \cdot \bar{x}_2 \cdot x_3 + \\ \ldots + \phi_{15} \cdot x_1 \cdot x_2 \cdot x_3$$

Grouping (ii): $\{x_0, x_1\}\{x_2, x_3\}$

$$F = (f_0^1) \cdot \bar{x}_0 \cdot \bar{x}_1 + (f_1^1) \cdot \bar{x}_0 \cdot x_1 + (f_2^1) \cdot x_0 \cdot \bar{x}_1 + \\ (f_3^1) \cdot x_0 \cdot x_1$$

where

$$(f_0^1) = \phi_0 \cdot \bar{x}_2 \cdot \bar{x}_3 + \phi_1 \cdot \bar{x}_2 \cdot x_3 + \phi_2 \cdot x_2 \cdot \bar{x}_3 + \phi_3 \cdot x_2 \cdot x_3$$

$$\vdots \qquad \vdots \qquad \vdots \qquad \vdots$$

$$(f_3^1) = \phi_{12} \cdot \bar{x}_2 \cdot \bar{x}_3 + \phi_{13} \cdot \bar{x}_2 \cdot x_3 + \phi_{14} \cdot x_2 \cdot \bar{x}_3 + \\ \phi_{15} \cdot x_2 \cdot x_3$$

Grouping (iii): $\{x_0, x_1, x_2\}\{x_3\}$

$$F = (f_0^2) \cdot \bar{x}_0 \cdot \bar{x}_1 \cdot \bar{x}_2 + (f_1^2) \cdot \bar{x}_0 \cdot \bar{x}_1 \cdot x_2 + \ldots \\ + (f_7^2) \cdot x_0 \cdot x_1 \cdot x_2$$

where

$$(f_0^2) = \phi_0 \cdot \bar{x}_3 + \phi_1 \cdot x_3 \\ (f_7^2) = \phi_{14} \cdot \bar{x}_3 + \phi_{15} \cdot x_3$$

Some of the results of this grouping technique are shown in Table 2, where they are compared with methods (a) and (b).

Table 2

	(i)	(ii)	(iii)	(a)	(b)
Number of levels	5	5	5	3	9
Maximum fan-in	8	4	8	16	2
Total number of gates	25	25	25	21	49

Evidently grouping (ii) may be preferable to either methods (a) or (b).

In general, the following may be taken as an optimal design strategy: if the fan-in is physically limited to g, the input variables should be divided into groups of h, where $2^h < g$.

CONTROL MEMORY ORGANISATION: FUNCTION SEARCHING

The memory of the universal digital learning element has as many single-bit storage locations as there are ϕ wires: that is, 2^n. The memory can operate in two ways. The first is a searching mode; that is, the memory content is changed when a "punishment" signal is received (such a signal indicates that the wrong function is being performed). The second mode, which is considered in the next section, involves the direct addressing of the bits in the memory.

The adaptive specification here is designed to cause the sequential memory to produce all the possible patterns at the control wires in response to an error signal at the input of the memory. On cancellation of the error signal the last found pattern at the ϕ wires is held. A binary counter where the error signal is a train of pulses will act in this way. The "punishment-reward" character of the identity information should be noted. If the instantaneous identity information output by the network is incorrect it merely calls for more pulses to the counter, and this continues until the correct function is found.

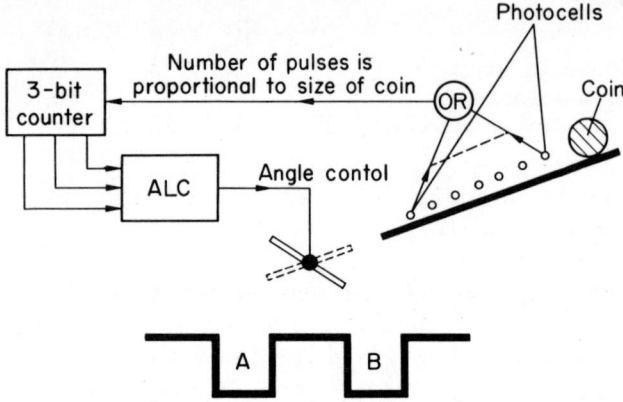

Fig. 6   Coin Sorter

To discuss the adaptation characteristics of this and other examples we consider the following application. A machine (Fig. 6) is required for sorting eight different denominations of coins into two categories. There are 256 possible ways of doing this and our circuit must be able to adapt to any one of these. The denomination is sensed by a set of photocells in a chute arranged so that the number of pulses from the system is proportional to the size of the coin. From 1 to 8 pulses can be emitted, this number being stored in a 3-bit counter. The learning element must perform the two-way classification on the output of this counter. Assuming that "training" trials consist of dropping coins down the chute and applying error signals until the right decision is made, it is evident that up to 256 training trials might have to be made in order to find the correct function. This is due to the fact that the system absorbs little information from the trials. The only advantage of this scheme is that the trainer can make errors, since the sequential memory will come round repeatedly to the desired pattern if it should be missed during the training procedure (after another maximum of 256 trials).

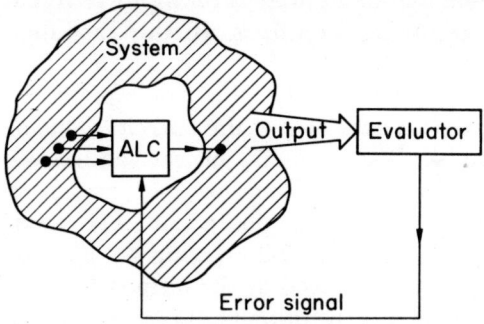

Fig. 7   Error-controlled ALC

Another application of a function searching element is shown in Fig. 7. Here, the element's performance is changed in order to improve the performance of the system as a whole. It is not necessary to measure the output of the element itself as is the case in some of the examples below.

BIT ADDRESSED MEMORY: SLAM ELEMENTS

SLAM stands for Stored Logic Adaptive Microcircuit, and refers to an element that is now being constructed in microcircuit form. Three-input devices are called SLAM-8s because they have 8 bits of memory, while four-input devices are called SLAM-16s. The latter has been manufactured as a 2 × 2 mm Metal–Oxide Silicon Chip. This device associates the applied identity bit with an n-bit input pattern by letting the pattern address the appropriate memory bit, which then absorbs the identity bit.

The block diagram of the device is shown in Fig. 8 and some logic circuit details are included in Fig. 9. The SI (Sensory Information) is applied at input terminals 1 to 4, terminal 4 being reserved for the Applied Identity

Information (AII), and terminals 1 to 3 for the PI (Pattern Information). The decoder causes each of the eight possible message patterns on input terminals 1–3 to excite one of the address lines. In this case the AII can only have two values, say 0 and 1. The excitation of an address line primes only the corresponding storage location so that it may accept the current value of the AII at terminal 4. The content of the addressed storage location is also transmitted to the output terminal. Thus, during a teaching phase the AII messages are associated with the PI, while during a "use" phase this associated II is reproduced at the output terminal even if it does not appear at terminal 4.

The clock terminal in Fig. 9 is included for technological reasons as it allows the user to choose the exact moment at

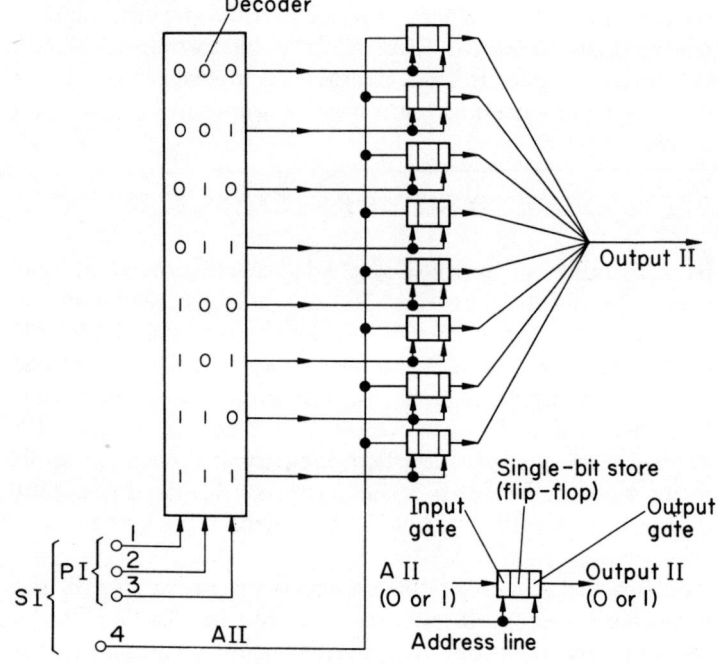

Fig. 8   Block diagram of SLAM-8

which the AII is to be associated with the PI. Terminal p allows SLAM devices to be coupled together to form larger SLAMS as shown in Fig. 10.

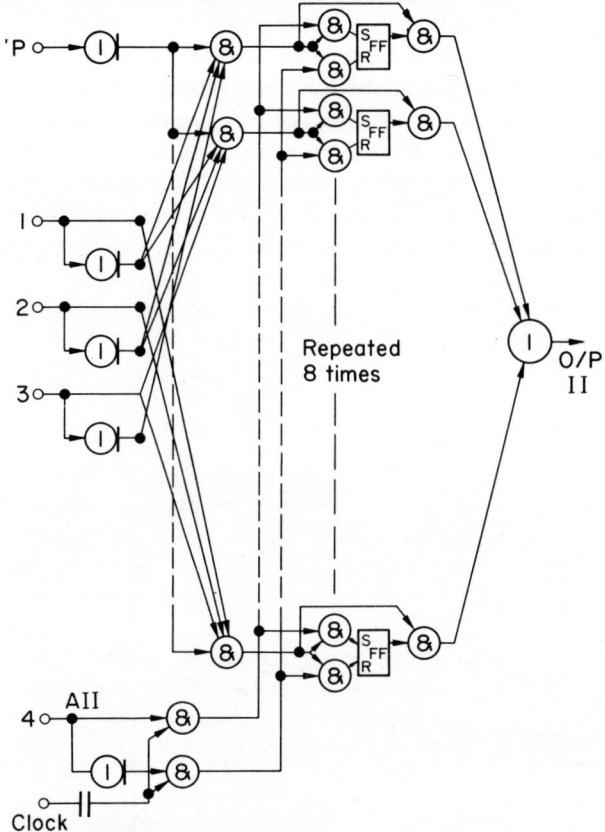

Fig. 9   Logic detail of SLAM-8

It is noted that in SLAM, the clear division between control memory and u.l.c. appears to have been lost. This is not strictly true, since all that has happened is that the u.l.c. has been "broken up" to allow it to address the memory.

SLAM type devices have been used as the basic building brick of Learning Computers and it is necessary, at this point, to give some reasons for choosing these devices rather than the more conventional threshold logic type of adaptive element such as an Adaline (Widrow and Hoff, 1961). The latter device evolved from models of neurons and may be impractical for the following reason. A real neuron has many thousands of synaptic inputs, and performs only a small percentage of the possible logic functions of these inputs. This probably accounts for some (but almost certainly not all) of the properties of generalisation of the brain (generalisation is the property of

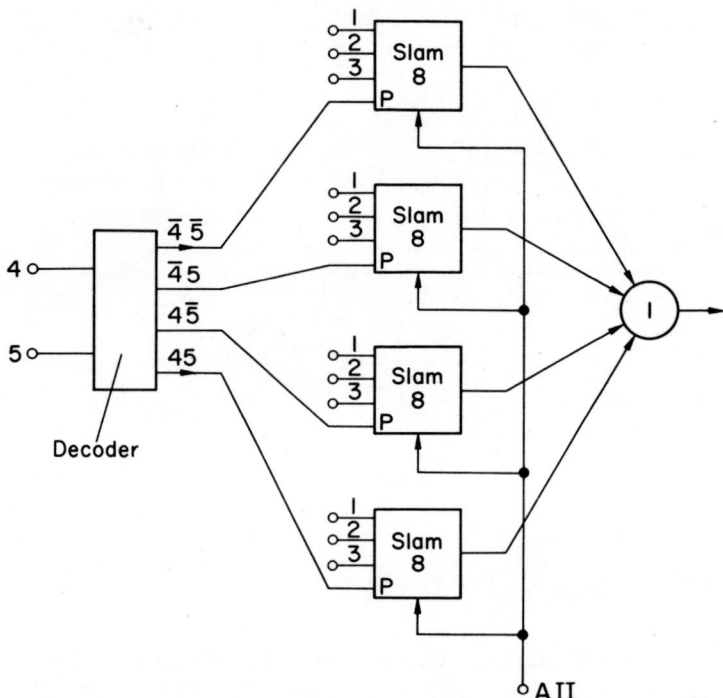

Fig. 10 Connection of Four SLAM-8 Elements to form a SLAM-32 (clock terminals not shown)

generating some reasonable II even for PI not seen during training—see next chapter).

Even with microcircuit technology it would be quite impractical to make a building brick containing thousands of digital counters (one for each synaptic weight). A scaled-down version with only a few inputs is clearly feasible, but restriction of the functions of a device with few inputs to only linearly separable ones (as those performed by an Adaline) is questionable. SLAM devices are universal in the sense that they can perform all the possible functions of their PI input terminals.

In the next chapter it will be seen that generalisation may be left as a property of the topology of an assembly of SLAM elements. This may be even more effective than if similar networks of threshold elements were to be used. Nevertheless, much of our description of the enhancement of the properties of learning nets through the addition of feedback (Chapter 5) does not depend on whether the elements are Adalines, SLAMs or perhaps real neurons.

3. Networks of Learning Elements

As indicated in Chapter 1, a single learning element does not possess the property of generalisation, that is, the property of assigning a reasonably correct output Identity to inputs not specifically seen during training. To show that an *assembly* of learning elements can generalise, we can consider a simple net of only two SLAM-8 elements as shown in Fig. 11. If one excludes the two AII terminals, it is seen that this net can receive 64 (i.e. $2^{(2+3)}$) PI messages and it can be taught to classify these into 4 (i.e. 2^2) II messages.

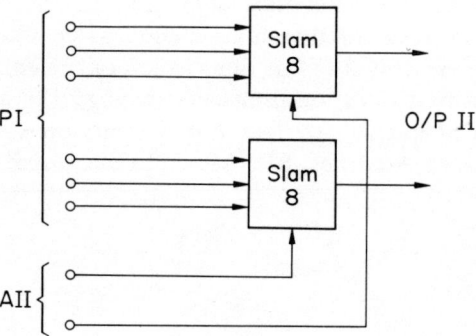

Fig. 11   A 2-element Single-layer Net (clock not shown)

Assuming that initially 0 is stored in all the locations, let us train the net so as to respond with $II = (11)$ to the input message (110 110) where the first 3 binary digits go to the first SLAM-8 and the rest to the second. Let us also train the net so as to provide the same response for (111 111). Each SLAM-8 now contains two 1's in its store and will therefore respond with $II = (11)$ not only to the two training patterns, but also to two additional PI patterns, namely (111 110) and (110 111). This indeed is a process of generalisation.

Much work has been done on larger versions of this net: single-layer nets. A single layer net has R inputs connected to R/k SLAM (2^k) elements. The signal on the R/k outputs is treated as a *response* in the sense that the number of 1s at the output is compared to the number of 0s. This generally assumes that the net has 0 in all storage locations before training. A training set of patterns is applied, for which the Applied Identity Information consists of a 1 at the AII terminals of each SLAM element. Subsequently, any excitation of the *pattern inputs* will cause a number of 1s to appear at the identity outputs. This number with respect to R/k is a measure of the similarity of the current input pattern to those in the training set. Generally, the inputs are connected *at random* to an input interface. In this context it is useful to define a generalisation factor, G.

THE GENERALISATION FACTOR G

G is defined as the ratio of the total existing input-output mappings of the net to the number of realisable ones. After training on some pattern set, a system with a high G will provide a "trained" response (of say, all 1's at the output assuming all storage locations to be initially at 0) for a set of patterns larger than a system with a low G. G is therefore a useful factor which may be expressed as follows. Consider a single-layer net with I inputs implemented with k-input SLAM devices. As the net has I/k outputs, the total existing number of mappings is:

$$2^{(I/k)2^I}$$

The number of mappings that can be realised by the net is $[2^{(\text{total storage})}]$ that is:

$$2^{(I/k)2^k}$$

Thus,
$$\log_2 G = (I/k)2^I - (I/k)2^k$$
which for $I \gg k$ is approximated by
$$\log_2 G \simeq I/k\, 2^I.$$

Table 3 shows the results of a simulation with a net where $I = 24$. The net was trained on 13 patterns. If *generalised set* is defined as the set of input patterns indistinguishably mapped by the trained net, r is the ratio of the order of this generalised set to the number of patterns in the training set itself.

Table 3

k	r	Cost (in bits)
24	1	16.8×10^6
12	14	8.3×10^3
6	430	256
4	1.5×10^3	96
3	2.6×10^3	64
2	32×10^3	48
1	300×10^3	48

However, there is a great deal more to generalisation than just the calculating of the order of the generalised set. Something must be said about the similarities of the patterns in the generalised set to those in the training set.

The random-connected single-layer net is a system which samples k-tuples of the input patterns. Thus, if the training patterns are similar to each other in the Hamming-distance sense*, the patterns in the generalised set must also be similar to those in the training set in the same Hamming-distance sense. This generalisation would certainly be labelled as "good" by an observer. In practice, the above situation occurs if the training patterns

* The Hamming distance between two patterns is the number of binary points in which the patterns differ.

are all slightly noisy versions of a prototype. The system after being trained on a few of these noisy patterns would recognise others.

However, if the training set consists of say upright and up-side-down versions of the printed numeral 4 it would classify patterns close to these prototypes but would almost certainly not recognise 4's rotated by 90°. An important maxim is encountered here which must apply to all combinational learning systems.

For generalisation to take place, a system must see a representative set of patterns in training. This probably applies even to neural brain nets. The additional properties of generalisation of the brain (such as recognising the 4 rotated by 90°) could be added to the net by making it sequential through the use of feedback, as will be discussed later.

A more accurate analysis involving Hamming distances now follows.

HAMMING DISTANCE ANALYSIS

It is assumed that the training process starts with all the SLAM stores set to 0 and that, for each pattern in the training set, an output of all 1s is impressed on the net by means of the AII terminals. Each training pattern is an R-dimensional binary vector, say t^j, where j is a distinguishing label for each of the T patterns of the training set. Thus $1 \leqslant j \leqslant T$. The test pattern in question is also an R-dimensional binary vector, say $\mathbf{x}$. If we perform an exclusive OR operation of $\mathbf{x}$ with t^j, element for element, we obtain a new vector $\mathbf{d}^j$. For example

$$\mathbf{x} = \begin{bmatrix} 0 \\ 0 \\ 1 \\ 0 \\ 1 \end{bmatrix} \text{ and } t^j = \begin{bmatrix} 0 \\ 1 \\ 1 \\ 1 \\ 1 \end{bmatrix}, \text{ then } \mathbf{d}^j = \begin{bmatrix} 0 \\ 1 \\ 0 \\ 1 \\ 0 \end{bmatrix}$$

The number of 1s in $\mathbf{d}^j$ is the Hamming distance H_j between $\mathbf{x}$ and $\mathbf{t}^j$. The nature of the SLAM element is such that, if n 0s of $\mathbf{d}^j$ occur at the k inputs of a particular SLAM device, that element responds to $\mathbf{x}$ with a 1 at its output. Therefore, if a particular SLAM element is to respond to $\mathbf{x}$ with a 1, the H_j 1s must occur outside the n inputs in question. There are $\binom{R-k}{H_j}$ such possibilities for a given H_j, the total number of possible patterns with H_j being $\binom{R}{H_j}$. Thus the probability of any one SLAM responding with an output 1 for pattern $\mathbf{x}$ considering only T_j is given by

$$P_j = \binom{R-k}{H_j} \bigg/ \binom{R}{H_j}$$
$$= \frac{(R-H_j)(R-H_j-1)\ldots(R-H_j-k+1)}{(R)(R-1)\ldots(R-k+1)}\cdot$$

One can now calculate the probability of a 1 being generated at the output of any one SLAM element after the occurrence of all T training patterns as

$$P_r = 1 - \prod_{j=1}^{T} Q_j \qquad (3)$$

where $Q_j = 1 - P_j$.

The probability of generating m 1s at the output of the net is then given by

$$\binom{v}{m} P_r^m Q_r^{v-m} \qquad (4)$$

where $Q_r = 1 - P_r$. The maximum likelihood is that $m = vP_r$.

Tests were carried out to compare practical and theoretical results on a laboratory microcircuit learning net that consists of 12 SLAM 8 elements (i.e. $k = 12$, $n = 3$, $R = 36$).

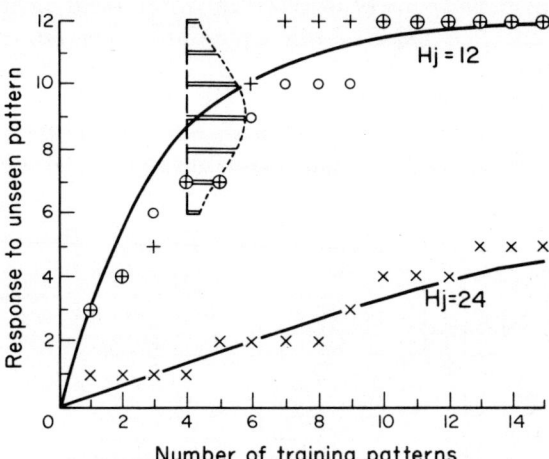

Fig. 12 Calculated and Measured Performance of a 12-element SLAM-8 Single-layer Net.
— Calculated Results
O Measured Results for Test 1, $H_j = 12$
+ Measured Results for Test 2, $H_j = 12$
× Measured Results for Test 3, $H_j = 24$

First, the net was trained on 15 patterns which all had a Hamming distance of 12 from a test pattern. The response of the net was calculated and measured for each value of T from 0 to 15. The results are shown in Fig. 12 (O). The calculated curves are those for the maximum likelihood values of m. The measurements were repeated for another set of training patterns with the same Hamming characteristics (+ in Fig. 12). A similar calculation was carried out for a set of 15 patterns with a Hamming distance of 25 from the test pattern. Measured results are shown as × in Fig. 12.

There appear to be some large discrepancies in the results obtained for the lower Hamming-distance sets. However, it must be remembered that the calculated curves represent the loci of the peaks of the probability-distribution curves, one of which is shown in histogram form in Fig. 12. The peak of this distribution corresponds to a probability of only 0·185, thus departures from the calculated curve are not entirely unexpected. Naturally, in this form of analysis, it is assumed all patterns with a particular H_j have equal probability of occurrence.

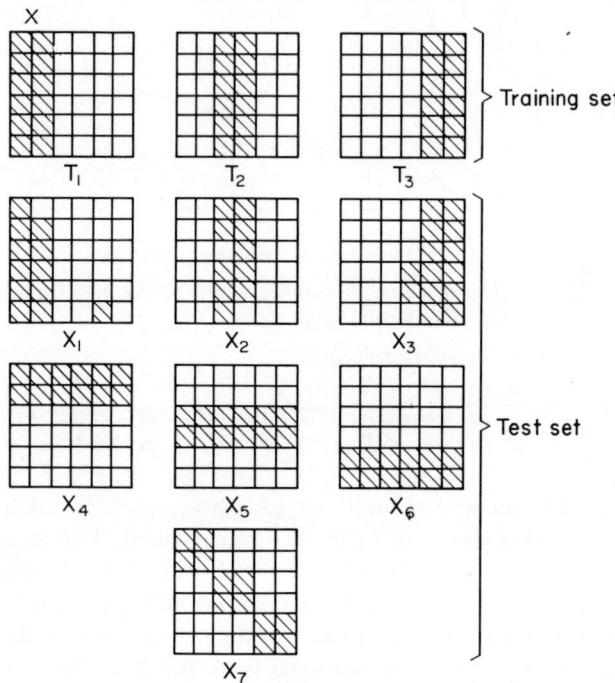

Fig. 13 Tests with widely different training patterns
T_1, T_2, T_3: Training Patterns
X_1 X_7: Test Patterns

One of the salient features of the single-layer net is illustrated by the patterns in Fig. 13. The net may be trained on patterns which differ widely in Hamming distance from one another. T_1, T_2 and T_3 in Fig. 13 are such patterns. Test patterns x_1, x_2 and x_3 are each close in Hamming distance to one pattern in the training set as in Table 4.

Table 4

	H_1	H_2	H_3
x_1	2	24	22
x_2	22	2	22
x_3	26	22	2

The low Q_j of the lowest H_j becomes dominant in Eqn. 3 yielding a high P_r. The calculated and measured results for these three patterns are shown in Table 5.

Table 5

	RESPONSE	
	CALCULATED	MEASURED
x_1	10·2	10
x_2	10·3	10
x_3	10·2	10

Patterns x_4, x_5, x_6 and x_7 are of interest because, in some way, they are combinations of the training patterns. They are each a Hamming distance of 16 from each of the training patterns. Thus, our theory predicts that the response peak is 4·85. The measured responses in fact are 3, 6, 5 and 6, respectively. This type of behaviour is important when the training set contains translated or rotated versions of a pattern, in which case a test pattern will be recognised if it is close in Hamming distance to at least one of the patterns in the training set. In fact, it is more likely that such a pattern will be recognised than any meaningless combination of the training patterns.

COMMENT ON OTHER TOPOLOGIES

Clearly, the single-layer net is not the only way in which learning elements may be connected. A general approach to the network theory of such elements is rather complex, and the reader is referred to the work on polyfunctional networks by R. H. Urbano (1968). Work on the cascading of single-layer nets is in progress at the moment and will be covered in future publications.

4. Applications

In Chapter 1 it was noted that a learning machine finds application in problems that are non-analytic in nature. The two fields in which most work has been done on this account are the recognition of visual patterns, handwriting in particular, and automatic control.

These are the two topics that are discussed in this chapter; they should be regarded only as examples of a much wider field of non-analytic data processing problems.

RECOGNITION OF HAND-WRITTEN NUMERALS

A network of SLAM-8 (stored-logic adaptable microcircuit with 8 bits of storage) devices is assumed. If the "retina" contains R sensors, $R/3$ SLAM-8 elements are connected to it. Each element has three inputs, so that each point of the retina is covered. This is done in a random, rather than an ordered, way, since it is initially assumed that any one of the possible 2^R patterns has an equal probability of appearing at the retina. Each adaptive element has one output, so that the whole net has $R/3$ outputs. The elements may be initially reset so that every pattern appearing at the retina produces a logic 0 at each of the $R/3$ outputs.

The net is "trained" by presenting to it a set of patterns for each of which a 1 is forced at all the outputs. Subsequently, the net is tested with other patterns, and if these belong to the same desired class as the training set (e.g. the training set may consist of handwritten 5s and the test set may also consist of other handwritten 5s), the number of 1s appearing at the output is a measure of generalisation. If, however, the test set consists of a different class (e.g. handwritten 6s), the number of 1s at the output is a measure of the error.

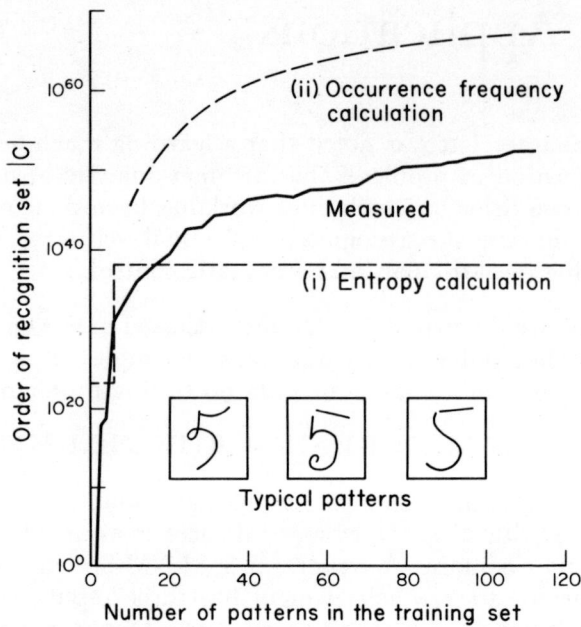

Fig. 14 Growth of the number of patterns equally classified, with size of Training Set

Each adaptive element processes the patterns appearing at three randomly selected points on the retina. We call these points a random 3-tuple feature. If for a particular set of training patterns, adaptive element i has "seen" s_i different 3-tuple features (where $1 \leqslant s_i \leqslant 8$), the total number of patterns that will produce a response of all 1s at the output of the net is $|C|$, where

$$|C| = \prod_{i=1}^{R/3} s_i$$

Clearly, if the patterns in the training set are similar to each other in the sense that they differ only in a few of their random features, $|C|$ will be small and will contain other patterns similar to those in the training set. By the same criterion, if the training patterns are different, $|C|$ will be large and will contain patterns that are not similar to those of the training set. Despite the latter undesirable characteristic, it is important to realise that the patterns in the training set always produce the right response.

To consider a realistic situation, we describe an experiment where the numeral 5 was written by a single person into an almost square box. The writer made these 5s as different as he could within his satisfaction that they were still 5s to a human observer. Typical examples are shown in the insert in Fig. 14. The box was formed into a 15×16 retina (i.e. $R = 240$) and a net-work of 80 SLAM-8 elements was simulated on a computer. The measured value of $|C|$ is shown in Fig. 14.

It is possible to predict this behaviour on the basis of the fact that the occurrence frequencies of 0s and 1s at the retinal sensors are likely to approach some stationary value for any set of patterns. To show this, we have calculated the entropy H of the handwritten patterns, where

$$H = \sum_{j=1}^{R} [f_j \log_2 (1/f_j) + (1 - f_j) \log_2 \{1/(1 - f_j)\}]$$

and where f_j is the frequency of occurrence of a 1 at retinal sensor j. Owing to the random connections, each adaptive element is sensitive to an entropy of $3H/R$ bits. A measure of the average number of different 3-tuple patterns seen by each element is $\lceil \bar{s}_i \rceil$, where

$$\bar{s}_i = 2^{3H/R}$$

and $\lceil x \rceil$ is the lowest integer greater than x. From this, the value of $|C|$ was calculated as shown in graph (i) of Fig. 14.

This graph confirms that the f_j values tend to become stationary as the size of the training set grows (this, in fact, is true even for very few patterns in the training set). However, the above is a probabilistic argument, whereas the network tends to act as an accurate integrator of the patterns that it sees. In other words, if the network were allowed to "forget" less-frequently-occurring patterns, the measured graph would approach the calculated graph (i) asymptotically.

A second calculation was carried out in which the integrating effect was taken into account. This was done by assuming that the f_j values were independent and by calculating the value of $\bar{s}_i$ as follows. Taking the average and most likely frequencies of occurrence of a logical 1 at the input of a SLAM-8 device as f_1, f_2, f_3, the probability of occurrence of pattern k, where k is an integer and $0 \leqslant k \leqslant 7$, is, say, p_k.

If pattern k is, say, $\{0, 1, 1\}$,

$$p_{011} = (1 - f_1)(f_2)(f_3).$$

The probability of pattern 011 having been seen after training with T patterns is Π_k, where

$$\Pi_k = 1 - (1 - p_k)^T \qquad (5a)$$

Then, on average, $\bar{s}_i$ is given by

$$\bar{s}_i = \sum_{k=0}^{7} \Pi_k \qquad (5b)$$

Graph (ii) shows the results of this calculation for the handwritten 5s. The difference between graph (ii) and the measured graph is an indication of the interdependence of the occurrence frequencies f_j. However, it appears that the rate at which $|C|$ grows is dependent on s_i, as calculated in

Eqn. 5(a,b). A further important result that has been obtained by calculations similar to the above is that graph (ii) depends heavily on the average entropy H, whereas it tends to be independent of the actual values of f_j. (The details of this calculation are beyond the scope of this monograph.) The conclusion can be drawn that the behaviour of the learning net is sensitive to the information content of the training patterns, as measured by H.

So far, we have dealt only with the size of the classified set of patterns generated by a set of training patterns. It is important to be able to assess the behaviour of the net with respect to unseen patterns where occurrence frequencies are known. If we let f_j be the occurrence frequencies of the training patterns, as before, and let f'_j be the frequencies of the unseen set, the probability of pattern k having occurred during training and in the unseen set is $\pi_k p'_k$, where π_k is calculated as before (for f_1, f_2, and f_3) and p'_k is calculated as p_k, but this time for f'_1, f'_2 and f'_3. The probability of a logical 1 being generated by the element in question is P, where

$$P = \sum_{k=0}^{7} \pi_k p'_k \qquad (6)$$

If, for a given net, an average value is obtained for the set $(f_1, f_2, f_3, f'_1, f'_2, f'_3)$, P may be taken as the likely response of the entire net.

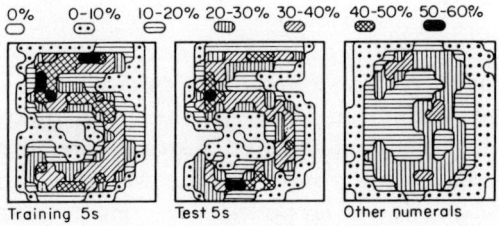

Fig. 15   Contours of Occurrence Frequencies

This is again illustrated with an experiment on handwritten numerals. Figure 15 shows the contours of the occurrence frequencies of a training set of 5s (calculated from 90 samples), an unseen test set of 5s (20 samples) and a test set of all the numerals except the 5 (110 samples of each numeral). It is significant that the contours for the 5s look like model 5s. The predominance of a 3-pattern in the contours of the other numerals is intriguing. In Fig. 16 we see that calculated and measured responses for this experiment are in fair agreement.

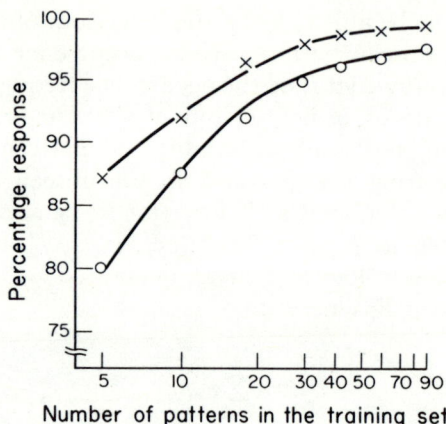

Fig. 16 Response of Learning Net
  × *Measured Response to 5's*
  O *Measured Response to other numerals*
  — *Calculated Response*

It is concluded that, in general, the occurrence frequencies of signals at the sensors of adaptive microcircuit learning nets provide important cues to the likely pattern recognition behaviour of the net. Furthermore, it appears that, on considering contours such as in Fig. 15 and applying Eqn. 6, it is possible to select nonrandom connections between the sensors and the net so as to obtain a much lower response for the patterns that should not be classified by the system. This is called a process of feature extraction to which we give a little more attention later.

Here, we show finally some results of the recognition behaviour of an 80-SLAM 8, single layer network (Fig. 17).

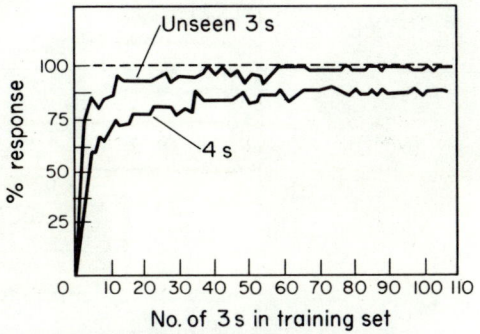

Fig. 17   Recognition of Hand-printed 3's and 4's by a network of 80 SLAM-8 Elements

This net was trained as hand-printed versions of the numeral 3 and tested against hand-printed 4's and unseen 3's. It is seen that generalisation works in such a way as to increase the response to the 4's as well as the 3's; however, after training on about 60 patterns 98–100 per cent response to 3 is indicated whereas no 4's were observed to give the same response.

A NOTE ON FEATURE EXTRACTION

Random connections in a single-layer net are used when one has no knowledge of the patterns for which the net is going to be used. However, after the occurrence of a training set, some knowledge (in the term of the SLAM memory content) has been acquired. It is possible to use this information in order to define a better method of connecting the net to a pattern interface. For example, it has already been mentioned that frequencies of occurrence give some information about such a connection.

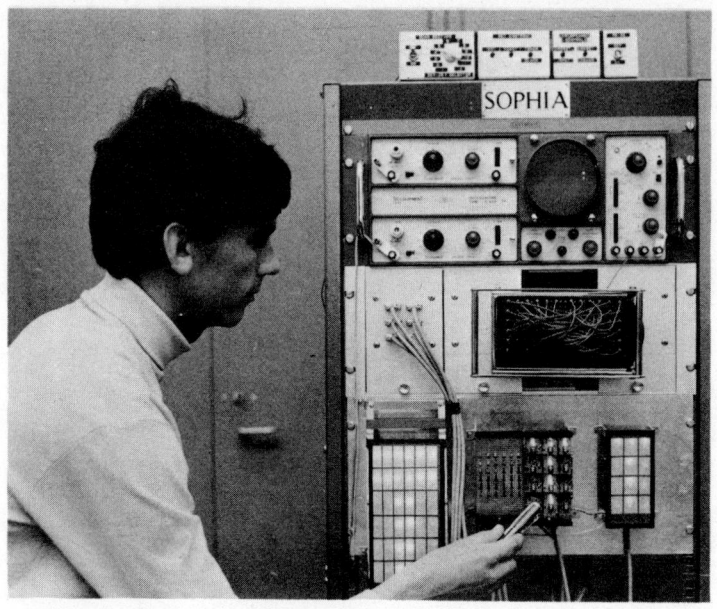

SOPHIA: A 12- (SLAM-8) Element Learning Computer

Much research effort is being devoted to this problem. For example, a favoured technique is to take note of the content of the store of all the SLAMs after training on two sets of patterns belonging to two classes between which the net is required to distinguish. The two stored contents are compared and SLAMs that contain similar stored patterns and reconnected. Early results with this technique have shown a considerable reduction in the recognition error, but much more experimental and theoretical work remains to be done.

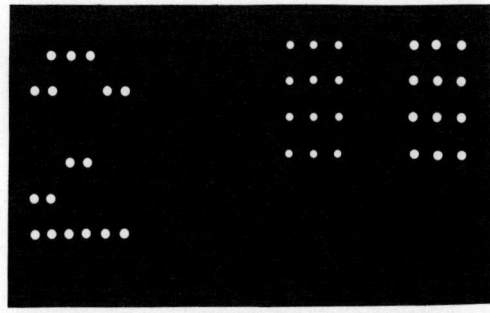

(a) Training SOPHIA to recognise a 2

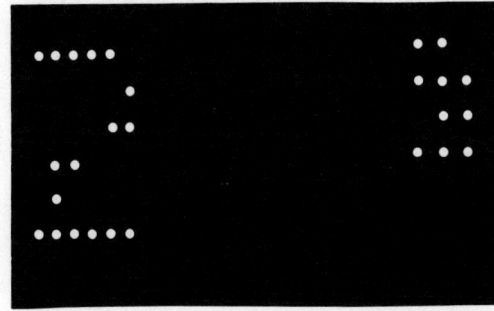

(b) Testing with another 2, a Response of 10/12 is obtained

(c) Testing with a 6, a Response of 4/12 is obtained

AUTOMATIC CONTROL

A typical way in which a learning machine may be used for automatic control is shown in Fig. 18.

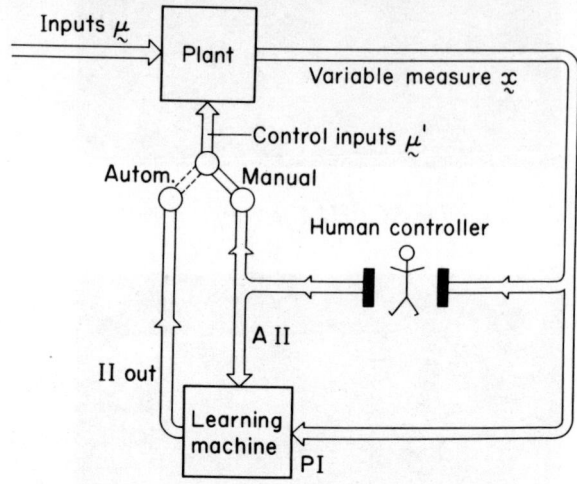

Fig. 18   Learning Controller with a Human Teacher

Here, the human controller supplies the AII to the learning machine while he is actually controlling the plant. This is the controlling information which must be associated with the parameter measurements.

Any dynamic plant may be represented by a set of differential equations (see Elgerd, 1967)

$$\dot{\mathbf{x}} = f(\mathbf{x}, \mathbf{u})$$

where $\mathbf{u}$ is a vector representing all the inputs to the system and $\mathbf{x}$ is an nth order vector representing measurements of plant variables. These are generally known as *state variables* and it may be shown that a set of controlling variables for the plant ($\mathbf{u}^1$) may be derived that are time independent functions of $\mathbf{x}_i$ that is

$$\mathbf{u}^1 = \mathbf{f}_c(\mathbf{x}, \mathbf{u})$$

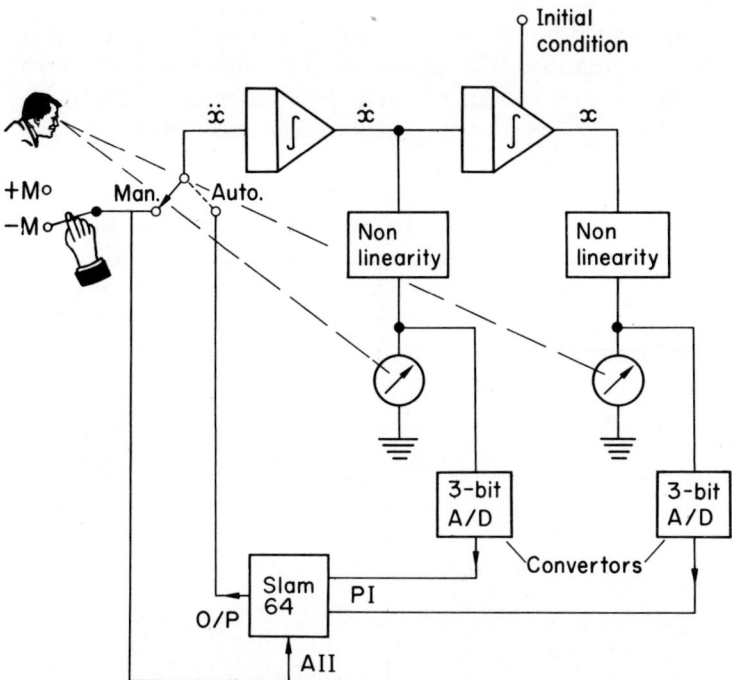

Fig. 19 Bang-bang control of a Second-order Plant with Non-linear Variable Measurements

In general, the slightest non-linearity makes the derivation of f_c very difficult if not impossible. If a human being learns to control the plant, the learning machine can absorb this control policy and thus effectively give us a solution of the unsolvable equation.

To illustrate what is involved, consider the simulated second order system in Fig. 19. Here, the state variables are measured non-linearly and a bang-bang (maximum effort) control may be supplied either by the human controller or the universal learning element. As long as the human can control the plant there is no difficulty in training the learning element, despite the non-linearities.

The system as shown has no generalisation, and much work remains to be done on the use of non-universal learning nets in this context. For some early results the reader is referred to the work of Thomas (1969).

It is often the case that all the state variables of a plant cannot be measured. In short, this means that the controller no longer performs a combinational task as the control function becomes time dependent. This means that a sequential learning system is required. Such systems are discussed in the next chapter.

5. Learning Nets with Feedback

A well-known characteristic of data processing machines where the output depends on sequences of input events (i.e. sequential machines) is that they can always be modelled as simple logic circuits (viz. combinational networks with no temporal behaviour) with feedback loops. These loops provide the temporal action, and the delays in the loops determine the timing of this action. Such circuits can generate long sequences of messages even if the input is held constant (see Aleksander, 1970c).

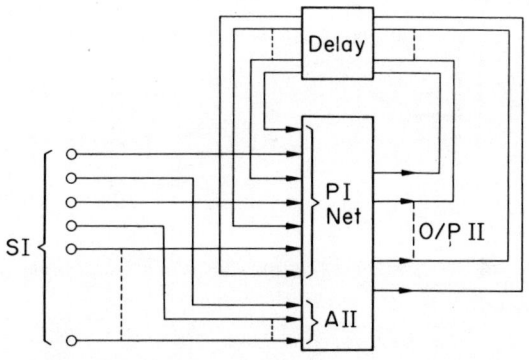

Fig. 20   Feedback in a Learning Net

The electronic learning nets discussed above, in common with learning nets developed by others, are generally combinational. It will now be shown that by the addition of a set of feedback channels with built-in delay, as shown in Fig. 20, sequential properties are added to the net.

The feedback connection is made between the output II and the pattern inputs PI. There seems to be little cause for having feedback to the AII terminals since what is fed back is merely a degraded version of what is needed at the AII terminals anyway.

The aspects of feedback nets considered in this chapter deal with the "intelligent" behaviour of the net and concern improved recognition, recall of sequences, sequence recognition and control of sequence generation ("attention"). These may be referred to as the *psychological* properties of feedback nets.

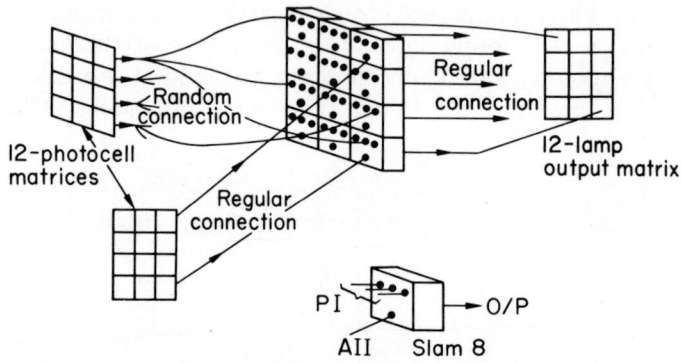

Fig. 21   A 12-element Learning Net with Overlapped Inputs

IMPROVED RECOGNITION

Experiments have been described by Aleksander and Mamdani (1968) in which a 9-element version of the 12-element net in Fig. 21 was used. In this case, the AII was supplied from a separate 9×9 matrix input and the 9 PI channels were connected to the input matrix only during training and to the output channels during testing.

Some of the results are given below:

SIMILARITY TEST
The net was trained to output the II:

```
    X X X
    . X .
    . X .
```

for the following set of PI:

```
X X X     . X X     X X X     X X X
. X .     . X .     . X .     . . .
. X .     . X .     . X X     . X .
```

The net was tested by applying a test pattern and connecting the feedback as soon as it responded. That is, the response to the test pattern was used as input, the new response being noted and used as a subsequent input.

For the following test pattern:

```
    X X X
    . X X
    . X .
```

the first response was:

```
    X X X
    . X .
    . X .
```

which, when fed back produced itself as a further response. We call this a process of *stable* pattern generation.

The test pattern:

$$\begin{matrix} X & X & X \\ . & X & . \\ X & X & X \end{matrix}$$

produced the output sequence:

$$\begin{matrix} X\,X\,X & & X\,X\,X & & X\,X\,X & & X\,X\,X \\ .\,.\,. & \to & .\,.\,. & \to & .\,X\,. & \to & .\,X\,. & \to \\ .\,X\,X & & .\,X\,. & & .\,X\,. & & .\,X\,. & \text{etc.} \end{matrix}$$

Finally, the pattern:

$$\begin{matrix} . & X & . \\ X & . & X \\ X & . & X \end{matrix}$$

gave rise to the sequence:

```
X . .     X X X     . . X
. X X → X . X → . . .
X . .     X X X     X . .


X . X     X X X     . . X
→ X X X → . . X → . X .
. X X     . X .     X X .


X X X     X X X     X X X
→ . X . → . X . → . X . →
. X X     . X .     . X .   etc.
```

One concludes that this arrangement may be described as a complex form of oscillation in which *recognition* is the formation of the stable II image at the output. The less similar the test pattern is (to those in the training set) the *longer* does it take to enter the stable oscillation. Indeed, more recent experiments have indicated that the likelihood of entering a repeated sequence unrelated to the original increases with dissimilarity between test and training patterns.

TWO-CLASS TESTS

Here the net was trained to output a pattern T:

$$\begin{matrix} X & X & X \\ . & X & . \\ . & X & . \end{matrix}$$

for the same input pattern T;

and to output pattern H:

$$\begin{matrix} X & . & X \\ X & X & X \\ X & . & X \end{matrix}$$

for the same input pattern H.

On subsequent testing within the pattern:

$$\begin{matrix} . & . & X \\ . & . & . \\ . & X & . \end{matrix}$$

the output sequence was:

```
X X .     X X X     X X X
. X .  →  . X .  →  X X .
. X X     . . .     . X .
```

```
          X X X     X X X
      →   . X .  →  . X .  →
          . X .     . X .   etc.
```

Testing with the pattern:

```
X . X
. X X
X . .
```

produced the sequence:

```
X . X     X . X     X . X
X X X  →  X X X  →  X X X  →
X X X     X . X     X . X   etc.
```

In other words, the first pattern with a Hamming distance of 3 from T and 7 from H causes the system to generate a stable T oscillation, while the second pattern is closer to H and is recognised as such.

The improvement in recognition lies mainly in the fact that recognition consists of the generation of a stable archetypal pattern. A net without feedback merely generates a *response* which has to be decoded into a recognition class. The feedback causes the decoding to be carried out automatically.

VARIABLE PRE-PROCESSING EXPERIMENTS
The patterns:

```
        X X X   X . .
        . X .   and X X X
        . X .   X . .
```

were made to generate a stable:

```
        X X X
        . X .
        . X .
```

This is analogous to teaching a human that ⊢ is also a T. The net subsequently recognises as T patterns those that are close in Hamming distance to ⊢ as the human would.

However, a net may be trained on:

```
X . X
. X .
X . X
```

to produce a stable version of itself in addition to its training on ⊢ and T, while also being trained to produce a stable:

```
. X .
X X X
. X .
```

This is analogous to the human having learnt that even though turning ⊢ "over in his mind" and recognising it as T is valid, turning × (multiply) "over in his mind" and recognising it as + (plus) is incorrect. This, in fact, is pre-processing of a kind which depends on reorganising the pattern before pre-processing takes place—a feat that is not generally achieved by computer-based preprocessing algorithms.

SHORT-TERM MEMORY

This is a property of the main feedback loop; it requires that the net be trained. Thus PI generates the appropriate II which, via the feedback loop, can sustain itself even if the PI is shut off. Examples of this have been seen in the last section. We note that the circulating II is generally a degraded version of the original PI. We experience this ourselves when we try to remember an object we have seen just prior to shutting our eyes. Our memorised image is a degraded version of the original—it contains less detail.

There is also bound to be some long-term degradation due to multiple passes of the II through the net. This is analogous to the fading of short-term memory. In an inexperienced net the circulating II would tend to fade more rapidly.

A typical example of fading in a partly experienced net is illustrated by results obtained with the net in Fig. 21 with feedback. The same input pattern was supplied to the AII and PI terminals. The net was trained to produce stable oscillations only for the following two patterns:

```
X X X  . . .
X X X  . . .
.  .  . X X X
.  .  . X X X
```

On testing with:

```
X X .
X X X
. . .
. . .
```

the sequence generated is:

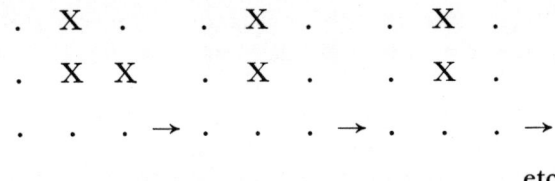

Whereas testing with:

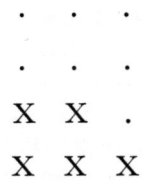

the sequence is:

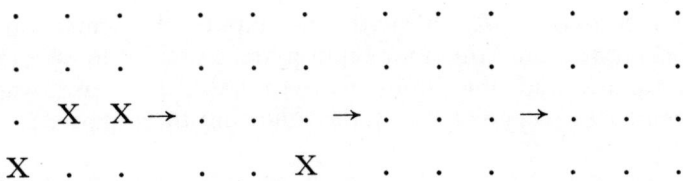

We see that the patterns on which the system has been trained essentially mean "top half ON" and "bottom half ON". Fading is due to the fact that the test pattern is not "recognised" entirely since the system has not been trained sufficiently. However, during fading the (bottom half)—(top half) characteristic of the pattern is retained.

RECOGNITION OF SEQUENCES

Let us assume that SI exists in "frames" and that a sequence of such frames has some meaning. Let us further assume that the delay in the feedback path precisely equals the timing between frames. First the net is trained to recognise a set of PI frames (say $f_1, f_2, f_3, \ldots f_n$) by producing stable oscillations for each frame as described in the section on "Improved Recognition".

During further training, SI is fed to the net in a sequence of frames, say $f_1-f_2-f_3- \ldots f_n$. An association will be formed between the II of frame f_t and the combination of the PI of frame f_t and the (fed-back) output II of frame f_{t-1} (which is additional PI according to Fig. 20). In a subsequent testing phase (AII cut off) recognition of the sequence takes the form of the II version of the sequence appearing at the output of the net if a similar sequence appears at the input.

To illustrate this, the following experiment was carried out on the 12 SLAM-8, machine (Fig. 21) with the feedback connections "OR-ed" with the input PI connection. During training, the same input pattern was fed to all AII terminals and the input PI terminals. The unit was trained to recognise stably the following three patterns:

```
X . .    . X .    . . X
X . .    . X .    . . X
X . .    . X .    . . X
X . .    . X .    . . X
  f₁       f₃       f₃
```

These patterns were fed to the net in the order:

$$f_1\text{--}f_2\text{--}f_3\text{--}f_1, \text{ etc.}$$

The AII for frame f_n was clocked into the circuit as soon as the output II frame f_{n-1} could be combined with the incoming PI of frame f_n. A check was kept on the contents of the stores of the SLAM devices, and after eight repetitions of the $f_1\text{--}f_2\text{--}f_3$ chain, only limit-cycle type changes were observed. This indicated that further training would be of no use. During testing, the AII was cut off and the output was observed as the input PI was presented in the order: $f_1\text{--}f_2\text{--}f_3\text{--}f_1$ etc.
The following II sequence was observed:

```
X . . .     . . .     . . . .
X . . .     . . .   . X X .
X . . →   . . . →  . X . .
X . .     X X .     . X . .
  (a)       (b)       (c)
```

```
. . X X .     . . . .
. . X X .   . X .
→ . . X → X . →  . X .  →back
. . X X .     . X .      to (d)
    (d)         (e)       (f)
```

65

It is seen that the output settles quickly to an almost perfect copy of the input delayed by one frame time.

The input frames were then applied to the net in a different order, namely,

$$f_1-f_2-f_3-f_1, \text{ etc.}$$

The first six outputs are shown below:

```
X . .     . . .     . . X
X . .     . . .     . . .
X . . →   . . . →   . . X
X . .     X X .     . . X
  (a)       (b)       (c)
```

```
. X .     X . .     X . .
. . X     . . .     X . .
. X X →   X X . →   X . X
. . X     . . .     . . .
  (d)       (e)       (f)
```

These were followed by a sequence of *eleven* relatively meaningless output patterns before returning to pattern (c) and repeating the 15 pattern cycle.

Thus, in the same way that the feedback net "recognises" a single frame of PI by generating a stable sequence of the pattern itself, it "recognises" a sequence of PI frames by reproducing a similar sequence at the output. This may well be a mechanism which takes place in humans. The recognition of a tune may consist of a sympathetic internal generation of the tune itself, the internal signals being the brain's output to the motor organs. That is, we may recognise tunes because we are humming them to ourselves.

RECALL OF SEQUENCES

A logic circuit with delayed feedback can produce long streams of output symbols when triggered by some input symbol. This is analogous to a person singing a song after hearing just the first few notes. The electronic net with delayed feedback loops may be taught to behave in this way. All that is necessary to memorise a sequence is that each frame of PI be associated with the II of the *next* SI frame.

As an example of this, the net in Fig. 2 (with feedback) was trained with the following patterns in order to cause it to generate a downward-moving bar pattern.

```
            PI          AII
(1)       X  X  X      .  .  .
          .  .  .      X  X  X
          .  .  .      .  .  .
          .  .  .      .  .  .


(2)       .  .  .      .  .  .
          X  X  X      .  .  .
          .  .  .      X  X  X
          .  .  .      .  .  .


(3)       .  .  .      .  .  .
          .  .  .      .  .  .
          X  X  X      .  .  .
          .  .  .      X  X  X


(4)       .  .  .      X  X  X
          .  .  .      .  .  .
          .  .  .      .  .  .
          X  X  X      .  .  .
```

After the connection of feedback, the first of the above patterns was fed to the machine and the following sequence of output II was generated.

```
      .  .   .  .  .   .  .  .
      X  .  X  .  .   .  .  .
      .  .   . → .  .  X →  .  .  .
      .  .   .  .  .   .  .  X  X

            X  .  X
            .  .  .
         →  .  .  .       back to
            .  .  .       first pattern
```

This is a slightly degenerate form of the desired sequence. When started with a vertical bar, the machine generated a meaningless sequence and went into a limit cycle with a meaningless pattern at its output.

Even though this determines that the net can "recall" sequences when primed by the first few (or just the first as above) frames, the training procedure is not a "natural" one. That is, one must determine whether the system can generate sequences after being trained solely by the application of SI sequences. To illustrate this, we must depart from the somewhat simple-minded OR combination of the PI and feedback II connections. In fact, we shall assume that the feedback II and input connections are connected in a disjoint fashion to the complete PI input of the net. The arrangement used in an experiment with the 12-element net is shown below:

Input Interface	PI Input of net	AII Input of net	Output II

B | A | B
 A | B | A
 B | A | B
 A | B | A
	C | A | C
	 A | C | A
	 C | A | C
	 A | C | A
		B | | B
		    | B |
		 B | | B
		    | B |
			C | | C
			    | C |
			 C | | C
			    | C |

A, Input PI connection.
B, Input II connection.
C, Feedback connection.

The experiment was designed to demonstrate the effect of training the net on the sequence:

```
X X X    . . .         . . .
X X X →  . . .    →    . . .
. . .    X X X         X X X
. . .    X X X         X X X
```

After training, the input:

```
X X X
X X X
. . .
. . .
```

was applied to the net and the following output sequence was generated (the input was removed as soon as the first output was obtained):

LEARNING NETS WITH FEEDBACK

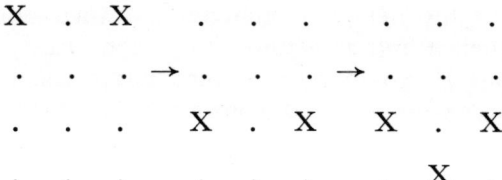

This is as close to the original sequence as the machine is likely to get.

ATTENTION

Attention may be defined as the ability of controlling recall and generation of sequences. There is no difficulty in controlling the output sequence of a non-learning sequential circuit. This is standard practice with devices such as reversible counters where the reversal is achieved by applying a control signal to an input terminal. The learning net itself may be taught to generate such signals in the output II so as to exercise control over its own generation of sequences.

To illustrate that this form of feedback is not an uncommon example of attention; consider the memory of a journey in a well-known town. Say that this proceeds along a road until a fork is reached. It is the memory of the fork (i.e. the output II) which generates the need for sequence control, since the memory sequence must proceed either to the left or to the right. One can decide either to suppress one of the possible routes, or perhaps to explore them both. Whatever the case, this decision is an example of sequence control generated by the ambiguous II of the fork.

A learning net can be taught to reproduce this behaviour. First, it must be taught to react to an ambiguous SI signal.

Second, if the net were sometimes trained on the left turn sequence and sometimes on a right turn one, the ambiguous II would be generated automatically whenever the net were required to generate the whole sequence autonomously. Even though this is a rather simple example of the concept of attention, it is rendered possible solely by the presence of the feedback loop and the application of the correct training procedure. This mechanism may be extended to more complex forms of attention without loss of generality.

For the time being, we must leave the possibility of complex attention merely as a speculation, since a net containing thousands of elements would be necessary to obtain sensible responses.

As a simple illustration, the following experiment was carried out on the 12-element net with OR-type feedback. The net was trained to oscillate between the two output patterns:

```
X  .  .        X  X  .
X  .  .        .  X  .
X  .  .   →    .  X  .
X  X  .        .  X  .
```

It was also trained to provide the following mapping:

```
X . X   . . X   . . X
X . X   . . X   . . X
X . X → . . X → . . X
X X X   . . X   . . X
```

and

```
X X X   . . X   . . X
. X X   . . X   . . X
. X X → . . X → . . X
. X X   . . X   . . X
```

The right-hand bar at the input was applied while the net was in oscillation and two output sequences were observed (one for each of the two oscillation states):

(a)
```
X . X   . . X
X . X   . . X
. . X → . . .
. . .   . . .
```

```
. . X   . . X
. . .   . . X
→ . . X → . . .
. . .   . . . etc.
```

and

(b)
```
    X  .  X  .  .  X
    .  .  X  .  .  .
    .  X  .  →  .  .  X
    .  X  .  X  .  .


             .  .  X  .  .  X
             .  .  X  .  .  .
       →  .  .  .  →  .  .  X
             .  .  .  .  .  .  etc.
```

This illustrates that the right-hand bar acts as a controlling input which stops one oscillation and leads into another.

COMMENT

The "psychological" behaviour of feedback learning nets has been emphasised in this chapter, but it should be remembered that such nets have other interesting properties. The ability to store long sequences of patterns which may be retreived by the application of some of these patterns at the input, provides the basis for the design of novel storage systems (see Prov. Brit. Pat. Appl. 33691/70). These have seductive associative properties, and will probably play an important role in future developments of computer hardware.

BIBLIOGRAPHY

On Digital Learning Computers

Aleksander, I. (1965), "Fused logic element which learns by example," *Electronics Letters*, **1**, p. 173.

Aleksander, I. (1966a), "Design of Universal logic circuits," *Electronics Letters*, **2**, p. 319.

Aleksander, I. (1966b), "Self-adaptive universal logic circuits," *Electronics Letters*, **2**, p. 321.

Albrow, R. C., et. al. (1967), "A universally adaptable monolithic logic module," *Electronic Communicator*, July/August.

Aleksander, I. (1967), "Adaptive systems of logic networks and binary memories," *Proc. Spring Joint Comp. Conf.*, p. 707.

Aleksander, I. and Albrow, R. C. (1968a), "Adaptive logic circuits," *Computer Journal*, **11**, pp. 65–71.

Aleksander, I. and Albrow, R. C. (1968b), "Pattern recognition with adaptive logic elements," *Proc. IEE–NPL Conf. on Pattern Rec.*, IEE pub. no. 42, pp. 1–10.

Aleksander, I. and Albrow, R. C. (1968c), "Microcircuit learning nets: some tests with hand-written numerals," *Electronics Letters*, **4**, p. 408.

Aleksander, I. and Mamdani, E. H. (1968), "Microcircuit learning nets: improved recognition by means of pattern feedback," *Electronics Letters*, **4**, p. 425.

Thomas, T. N. (1969), "Adaptive Logic Circuits for automatic control," Ph.D. Thesis in Engineering University of London.

Aleksander, I. (1970a), "Brain cell to microcircuit," *Electronics and Power*, **16**, pp. 48–51.

Aleksander, I. (1970b), "Some psychological properties of digital learning nets," *Int. Journ. Man-Machine Studies*, **2**, pp. 189–212.

Aleksander, I. (1970d), "Microcircuit learning nets: Hamming distance behaviour," *Electronics Letters*, **6**, p. 134.

Aleksander, I. and Fairhurst, M. C. (1970), "Pattern learning in humans and electronic learning nets," *Electronics Letters*, **6**, p. 518.

On Learning Systems in General

Widrow, B. and Hoff, M. E. (1961), "Adaptive switching circuits," *I.R.E. Wescon Convention Record.*

Nilsson, N. J. (1965), "Learning machines," *McGraw-Hill Systems Science Series.*

Michie, D. (1967 etc.), "Machine Intelligence," Annual series—Oliver and Boyd.

Textbooks

Elgerd, O. (1967), "Control Systems Theory," McGraw-Hill.

Aleksander, I. (1970c), "Introduction to logic circuit theory," Harrap.

INDEX

Adaline	30
Attention	59, 71–74
Automatic control	50–52
decoding	60
Behaviour	12
combinational	12
sequential	12
Bit-addressed memory	27
Broom balancing	14
Coin sorting	26
Combinational behaviour	12
logic	53
Control, automatic	50–52
memory	19, 25
Controller	50–52
human	51
Decoding, automatic	60
Entropy	43
Extraction of features	47, 48
Feature, extraction	47, 48
three-tuple	42
Feedback	16, 53–74
Function searching	25
Generalization	12, 32
factor	33
Hamming distance	35–39
Hand-written numerals	41
Identity	19
channels	12
information	27, 31
Learning, defin. of	9, 10, 11
element	18
Logic, combinational	53
sequential	53
universal	20–25
Memory, bit-addressed	27
control	19, 25
short-term	61
Network, of SLAMs	39
topology	39
Neurons	11, 18, 30
Numerals, hand-printed	41
Oscillations, stable	57
Pattern, channels	15
information	28, 31
recognition	15, 16, 41–49
sequences	56–60
Patterns	41–49
Preprocessing, variable	60–63
Psychological properties	54
Recall, of sequences	67–71
Resistor, variable	9–11
Recognition, improved	54–61
of patterns	15, 16, 41–44, 57
of sequences	64–67
Searching, function	4, 25
Sensory channels	12
Sequence, recall	67–71
recognition	64–67
Sequences of patterns	56–60
Sequential behaviour	12
Sequential logic	53
Similarity tests	55
SLAM	19, 27–30
network	39
Sorting, coin machine	26
State oscillations	57
State variables	16, 50, 51
Tests, two-class	58
Three-tuple feature	42
Thought	16
Topology, of network	39
Training	13
Two-class tests	58
Universal logic	20–25